Crafts in Plastic Foam

Crafts in Plastic Foam

Alan Barnsley

Watson-Guptill Limited   New York

First published in the United States 1973
by Watson-Guptill Publications
a division of Billboard Publications Inc
165 West 46 Street, New York, NY

Copyright © 1973 by Alan Barnsley
First published 1973 in Great Britain by
B T Batsford Limited
4 Fitzhardinge Street, London W1H 0AH

All rights reserved. No part of this
publication may be reproduced or used
in any form or by any means—graphic,
electronic, or mechanical, including
photocopying, recording, taping, or
information storage and retrieval systems—
without written permission of the publisher.

Manufactured in Great Britain

Library of Congress Cataloging in
Publication Data

Barnsley, Alan
Crafts in plastic foam

SUMMARY: Instructions for creating a
variety of decorative objects from
polystyrene materials.
Bibliography: p.
1 Plastics craft.   2   Polystyrene.
[1   Polystyrene.   2   Plastics craft]
I   Title.
TT297.B37 1973   745.57   72–11573
ISBN 0–8230–0999–8

Contents

Acknowledgment		6
Introduction		7
1	Expanded polystyrene	8
	Equipment and materials	9
	Making a hot-wire cutter	12
2	Take a tile . . .	14
	Printing	14
	Cutting and design exercises	17
	Movement towards three dimensions	21
	Jigsaws	32
	Mosaics	33
	Action of solvents	35
	Applied work	41
3	Take a slab . . .	44
	Melted wax reliefs	53
	Carving and casting	56
4	Take a block . . .	62
	Large carvings	62
5	Take some scrap . . .	81
Further reading		93
Suppliers in Great Britain		95
Suppliers in USA		96

Acknowledgment

For permission to use the many photographs of work from pupils at my school, my thanks are due to the Director of Education for Lindsey, the governors of the school and the headmaster. Also to Mr David Freeman, adviser for art and design for Lindsey and Mr J. V. Starkey, my assistant at school, for their great help and co-operation.

Many of the photographs have been loaned to me by professional artists and I extend my thanks especially to Vencel Ltd, Mr Kiss of Design and Display, and the Cement and Concrete Research Association; also to William Mitchell, Keith McCarter and Derek Howarth.

Much of the work of my pupils has been made possible by the generosity of the commercial concerns in the list of suppliers.

Finally, I would like to thank my wife, Jenny, for her help and patience in allowing our house to become a polystyrene store and display centre.

A.B.

Cleethorpes 1973

My thanks for the loan of photographs are due to:
Vencel Ltd, figures 106, 112, 113
Design and Display, figures 81, 113–118
Cement and Concrete Association, figures 82–88
Super Tools Ltd, figure 3
William Mitchell, figure 93
Derek Howarth, figures 108, 109
John Vickers, figure 107

Introduction

Necessity is the mother of invention. A very old phrase but often the maxim adopted by teachers, because of limited stock allowances in schools. All around us we see materials which can be adapted for uses other than those for which they were originally intended. Children, if they are given the encouragement and opportunity can find great excitement and satisfaction in creating things from simple materials which would normally be destined for the scrap heap.

To be added to the already extensive list of art room *found* materials, is expanded polystyrene. A mere feel of its substance and the look of its crystalline whiteness whets a child's curiosity and appetite for experiment. The ability to balance a large cube on one hand is the start to the fascination of expanded polystyrene as an art form. Its main advantage over many traditional materials is its range and versatility. It is suitable for work ranging from that of an infant to the complicated carvings by professionals.

It offers limitless potential both aesthetically and industrially, and its uses range from the composition of mosaics, sculpture and commercial design to the casting of concrete fascias, packing and the insulation of buildings.

In school, its main value lies in the ease with which it is worked; also in that it involves very little expense or equipment. Three-dimensional expression is brought within the range of all pupils as this material does not require any physical strength. Children in remedial classes who lack concentration find renewed interest and enthusiasm in seeing their work being created so quickly, and they are able to work on a large scale.

On the following pages, various processes and possibilities are described which provide a basis for further creative development and show the potential of this plastic material.

If a child has no chance to create, he will destroy. He needs to be able to create with materials that challenge the spirit of adventure and inventiveness, and to be able to build on what has been discovered. Expanded polystyrene offers these opportunities.

1 Expanded polystyrene

Expanded polystyrene belongs to the family of *thermoplastics,* which means that it softens when heated and hardens while cooling. This is the key to most of the processes that can be used for our purposes. Its open cellular structure presents some problems for carving; some traditional tools, such as those used for stone carving which work by pressure, tend to make the cells collapse or crumble. The most successful method is to melt the plastic foam apart. This is done by applying either hot-wire cutters or heated carvers. Tiles can be cut cleanly by using a very sharp knife or a safety razor blade.

One valuable property, especially for casting, is that the foam is soluble in solvents. Many solvents can be very harmful and work must be under the close supervision and control of the teacher. Harmful vapours or gases may be produced through the evaporation of solvents, and it is necessary to ensure that their concentrations are kept as low as possible. *Work in a well-ventilated area.*

The main advantages of expanded polystyrene

It has a very high strength to weight ratio.
It is durable and resistant to moisture.
It is low in cost, readily available, easily fabricated.
Burning rate: slow or self-extinguishing and it complies with the most stringent of fire regulations.
It has a high standard of finish and accuracy.
It is cut and carved, which means that it does not require physical strength, nor expensive apparatus.
Almost any finish or texture can be created and traditional materials simulated.
Large work can be quickly executed, easily displayed and mounted.
It is very versatile and adaptable and has tremendous potential and scope for experiment.
Its lightness, 'warmth' and whiteness arouse instant interest and curiosity.
It is clean to work with and therefore suitable for use in non-specialist rooms.

Expanded polystyrene is easily obtainable from builders' merchants, some art suppliers, or direct from specialist commercial concerns. Most children are able to find sources of broken or once-used ceiling tiles and a local insulation factory may be willing to donate off-cuts usually up to 100 mm (4 in.) in thickness. Our most exciting collection has been in the form of packing from the electronics industry. These are extremely good for abstract prints and for the building of 'junk' sculpture.

Many plastics require special protection because of the risks to health. Expanded polystyrene is a safe material, inert and non-toxic, and there are no hazards from oral ingestion or skin contact. The application of heat with a hot wire, however, does release styrene fumes, which in large quantities can be dangerous. This should be carried out only in well ventilated conditions so that a low concentration of vapour is obtained. The ventilation should be sufficient to maintain a supply of fresh air in the work area at all times. The hot wire should be operated at a temperature below red heat.

The smaller cell expanded polystyrene, often referred to as *Styrofoam*, requires a generous supply of fresh air in the work area.

Expanded polystyrene, unless specifically fire-resistant, is highly inflammable and must be stored away from open flames or other heat sources.

Equipment and materials

As mentioned earlier, one of the advantages of expanded polystyrene is that it does not require expensive apparatus. Most of the equipment can be improvised on a very small budget.

Carving and cutting tools

Tiles can be cut cleanly and easily using a very sharp knife or a one-edged safety razorblade. Large blocks of polystyrene can be cut with traditional saws, and shaped with files, rasps, metal all-purpose planes and wire brushes.

Hot-wire cutters

These work on the principle of an electrical current passed through a nickel-chrome wire. The heat produced cuts the material. A small battery-operated, plastic hand hot-wire cutter can also be used.

Heated carvers

A heated carving tool can be purchased in the form of an electric soldering iron, supplied with a variety of wire attachments shaped to give different carved forms on the polystyrene. However, packing case wire, mounted in wooden handles, produces satisfactory results. For detailed, intricate work, a needle embedded in a cork, or clipped in a clothes peg, has produced excellent results.

Heat

For younger children, the safest supply of heat for the carvers is either candles or night lights. For older children, a bunsen burner or a gas ring may be used. At no time should children be left unsupervised whilst working with naked flames or gas burners. Check the school fire regulations and have an extinguisher handy.

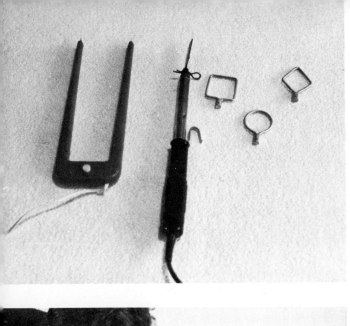
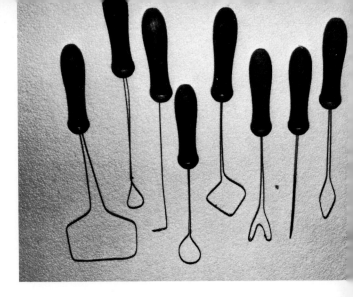
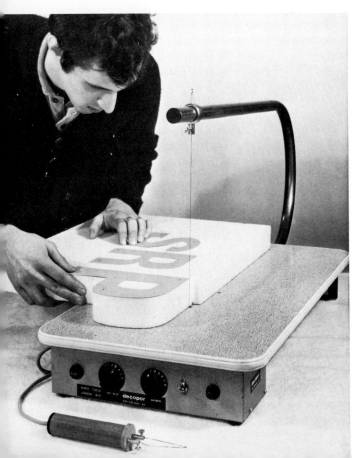

Joining and gluing

It is possible to join pieces of expanded polystyrene by passing a hot-wire between their surfaces, and pressing them together. The same method can be used with the melting effect of a solvent. However, these two methods have never proved satisfactory to me, and I much prefer to use the adhesives specially produced for plastic foams. Most adhesives are suitable as long as they are of the non-solvent variety, but it is advisable to experiment. Heavy-duty wallpaper paste is economical for use with tiles. For larger work I have found that adhesives with a polyvinyl acetate emulsion base satisfy all requirements, such as *Marvin Medium, Tretobond, PVA, Sobo* and *Elmer's glue*. Plaster also makes strong joints. It is always advisable to have a supply of dressmaking pins or wooden cocktail sticks available, these hold the pieces together while the adhesives set.

1   Hand operated battery hot-wire cutter and electric soldering iron with a variety of shaped carving attachments
2   Shaped heat carvers made from packing case wire
3   An excellent hot-wire cutter, with adjustable voltage and incorporating an electric modelling tool. To produce accurate letters, use cardboard stencils attached to the polystyrene with pins

Making a hot-wire cutter

A hot-wire cutter is an implement that holds a heated nickel-chrome wire taut and melts expanded polystyrene when it is in contact with it. It is simply constructed, especially with the help and co-operation of the school metalwork, woodwork and science departments. The cutting wire has to withstand a high temperature and for this nickel-chrome is used, which can be continually heated and cooled with little or no loss of strength.

The cutter table consists of a 50 mm × 25 mm (2 in. × 1 in.) timber framework with a 12 mm ($\frac{1}{2}$ in.) thick board, with a formica surface to give a smooth working area. A mild steel rod is bent into a U-shape, to support the cutting wire. One end of the rod is fastened beneath the table, near the centre, through a hole in the side frame. The nickel-chrome wire passes through a 7 mm ($\frac{1}{4}$ in.) hole in the table top and is connected underneath to an electrical contact, the other end to the U-shaped rod. Normal domestic voltage is connected to a variable transformer ($4\frac{1}{2}$ volts produce enough heat) which can be built into the machine. An on/off switch is a useful extra. An earth must always be provided. The alligator clip produces a variable resistance, finely adjusting the voltage to accommodate varying thicknesses of polystyrene. The hot-wire cutter *must never be connected directly* to the main electrical supply; a transformer must be used.

To improve upon the basic design, and to make a super de-luxe model, the table and sides are covered with a plastic laminate.

This cutter is good for freehand cutting, but for precision cutting and circular and cylindrical shapes, further experimentation is necessary. For cutting circles, a series of small holes are drilled in the laminate surface, radiating from the wire. A steel pin, or compass point, is pushed through the polystyrene into the required hole, and by revolving the material against the hot wire, one can produce a circle or cylinder.

To produce parallel and regular shapes, an adjustable sliding fence is fastened to the side frame. Usually the U rod is at right angles to the cutting table. One further refinement is to loosen the clip fastening the U rod underneath the table, move the U rod at an angle to the cutting table, and then retighten the clip. The hot-wire will now be at an angle to the cutting table, and this adjustment will produce angled cuts. This helps in the making of shapes for the technical drawing department.

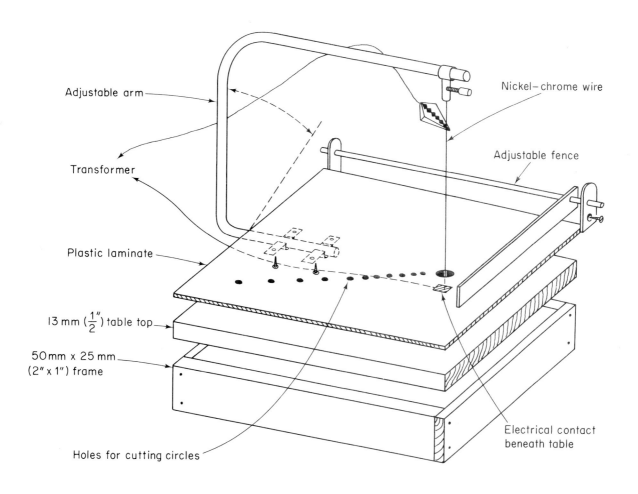

4 To make a hot-wire cutter

2 Take a tile...

Printing

The soft surface of tiles can lead to the creation of different surface textures and the making of exciting prints. A variety of objects are pressed into the surface and the indentations produce rich decorations. These can be shown to their best advantage as inked prints. The variety of prints produced and 'found' objects used, are limitless and could serve as an introduction to printing, suitable for pupils of all ages.

By drawing heavily on the surface of a tile, a line block can be produced. An extension of this method is the use of heat for producing various patterns. The blocks are subsequently inked and prints taken.

The tip of a pin fixed in a cork or clothes peg, heated in a candle flame, can be used to etch lines into the surface of a tile. Tiles can be textured for patterned prints. Water or oil-based printing inks may be used, but some cellulose-based printing inks do react against the polystyrene tiles. They tend to dissolve the surface and produce a stippled effect on the prints, though this may still be used to advantage.

5 Expanded polystyrene tiles have a surface texture that is worth exploring. Using a roller and printing ink, the cell structure of an ordinary flat tile is revealed on the print

6, 7 Some commercial tiles have deliberate decorative patterns. These can be successfully used for collage prints

8 Shaped heat cutters can be used to produce textured tiles. The tile shown has a regular pattern. From this example only, it can be seen that there are many possibilities. Children love to experiment and this exercise is of great value in assessing the amount of heat required, the depth and shape of cut, and the care needed to handle and control the heated cutter

9 This tile was decorated by pressing with the thumb into the surface; it was then lightly covered with an inked roller

10 A variety of bottle tops have been pressed into the tile. This exercise was limited to the use of objects with the same shape

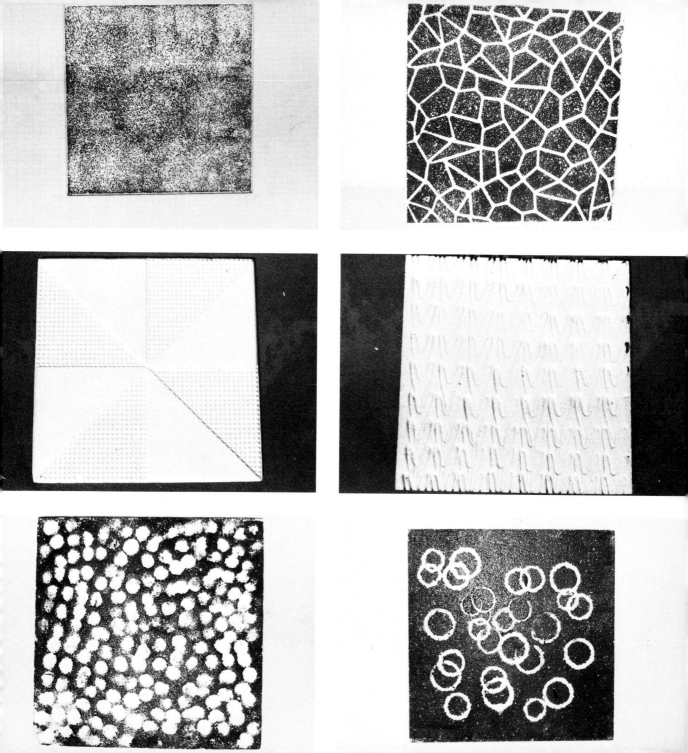

With a monochrome (single colour) print, the excitement of the design is dependent upon the balance between solid areas of colour and white, with the careful addition of textures.

The photographs show some of the possibilities of expanded polystyrene as a medium for printing, from single to multi-coloured prints. It may be found that the surface of certain tiles have too open a cell structure. A very smooth surface, excellent for printing, can be found on the reverse side of the special, commercial decorative tiles or the small cell structure of the type of polystyrene used for packaging.

11 A group exercise by eleven-year-old children, using a candle and heated needle. Each tile shows an individuality of style and great variation in design

12 Girls aged eleven have made this selection of tiles that have been used for printing. The ease and speed of producing these prints helps to retain interest. Young children are quickly able to understand the principles of printing and do not have to overcome the physical difficulties encountered when using linoleum as a printing material

13 Spaceman *(boy aged 12)*. The open surface of the tile has been used advantageously in this print. The subject is enhanced by a speckled background and a not too solid print

PLATE 1 (opposite) Detail of a mosaic 360 cm × 121 cm (12 ft × 4 ft)

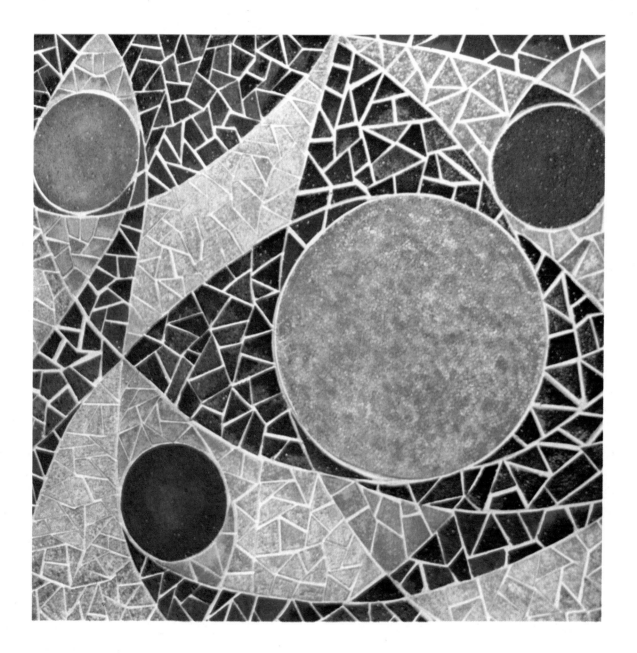

Cutting and design exercises

Expanded polystyrene tiles can be cut cleanly and easily using a sharp craft knife or a safety razorblade. The following examples have been produced with knives or a battery operated hot-wire cutter.

14, 15 This is a useful exercise in getting to know the feel of the material. A pleasing design is the result of a relationship and rhythmical balance of shapes. Every cut must serve a purpose and must be viewed in relationship to the whole tile

16   Simple cutting exercises can be extended. The 'snowflakes' can be easily transformed into Christmas mobiles and it can be seen that the concept of negative and positive forms has been created by experimenting

17   Coloured cellophane glued to the reverse of the tile has transformed a simple exercise into an easily produced stained glass window. Again, this is suitable for Christmas decorations
   Do not throw the cut-out pieces away. They will come in useful later!

18 The shapes have been cut cleanly through the tile. The pieces are removed from the tile, and their edges and those of the tile are filed away at an angle. The pieces are then put back in the tile in their original positions and glued, giving a sculptural effect. A textured decorative tile has been used in this example, and the surface contrasts with the smooth, filed edges
19, 20 From a flat tile, and with the limitation that nothing must be added or taken away, a relief tile is produced. The cut-out shapes are moved on the remaining tile surface until they look correct, and are then glued. The shapes are all related, and by experimenting, the discovery of general design principles is developed.

20

Movement towards three dimensions

A relief relies upon the correct and careful control of light and shade. This must be a consideration throughout the exercises, so that the material and product may be shown to its best advantage.

21–23 Although children are allowed much freedom to experiment, certain guide lines are suggested. This exercise is confined to simple cutting, nothing added or subtracted, and negative and positive forms. Natural forms, a tree shape and a strawberry, suggest the design. One half of the tile is cut and the cut piece is turned over and glued to the other half of the tile

Throughout the exercises, an understanding of the material is discovered by practical experience.

Many fundamental principles of design and the forming of shape relationships apply to all forms of art communication, and through the use of expanded polystyrene, can be transferred to other media.

The following pages are expansions to the basic theme and limitations suggested. They require little explanation and are only a small sample of a vast number of possibilities. The examples used have been produced by children aged twelve years.

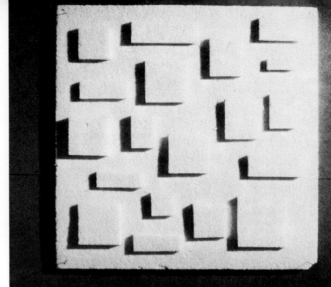

24 The shapes have been cut through the tile, eased up from below, and glued in the raised position
25, 26 A positive and negative view of the same tile as a result of pushing cut pieces upwards or downwards

27, 28 An extension of cutting and rearranging. Each rearranged piece is further subdivided to produce a pleasing whole. Careful observation is required to see how these reliefs were formed

29, 30 Positive and negative. Reliefs with five planes are produced by several cuts in each shape, which are then pushed upwards or downwards

31 This hexagonal design has incorporated many of the previous exercises and has produced an attractive honeycomb design

32, 33 Without adding or taking away anything, cut pieces of a tile are turned at right angles to the surface of the tile. The different sizes of the pieces produce different heights of relief and shadow. A controlled use of light enhances the relief and emphasises light and shade

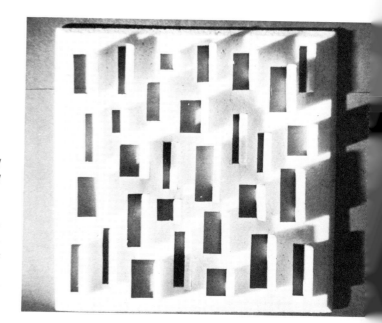

34 The square shape of a tile is transformed by cutting and by slight movement outwards. The work can be displayed, or rearranged, by fixing the pieces with steel pins into a sheet of paperboard or thick card, without having to use glue. There are many variations on this theme

35 Tiles, unlike cardboard, have a noticeable thickness, and therefore are not restricted to two dimensions. The pieces of tile have been overlapped and have produced a relief tile. This principle could be applied in making contour maps in relief

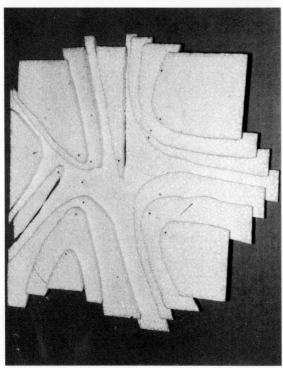

36 The same pieces have been glued at right angles to the original tile. This introduces children to freestanding three dimensional work, using pieces they already know

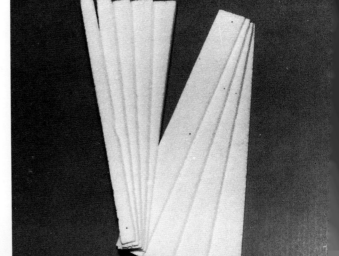

37–39 A series of parallel vertical cuts can produce a sense of movement, especially by decreasing the width of each strip. This sense of movement can then be translated into three dimensions

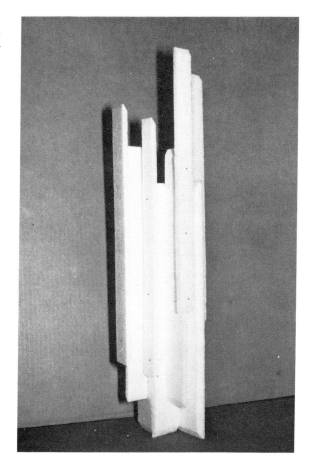

40 The same strips of tile can then be used as a form of unit construction. The pieces are first fastened together with steel pins. When a pleasing construction has been made, the pieces are glued together

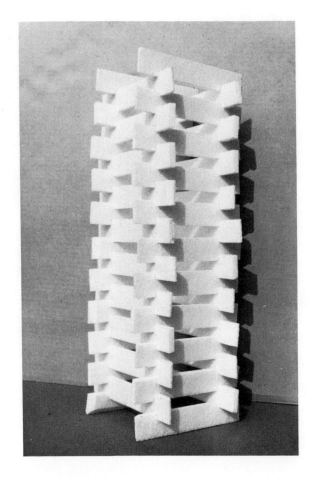

Apart from being a basic exercise in three dimensional construction, these structures can serve as maquettes for larger work in stone slabs or timber.

41 Strips of tile, all the same size, are used in this construction. One limitation has been imposed. The structures are to be formed using interlocking joints, without resorting to pins or adhesive

 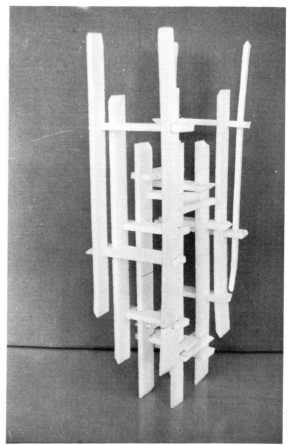

42, 43 Pieces of tile, the same width but with differing lengths, form these constructions. All the joints are interlocking and no glue has been used. It was found necessary to use steel pins in the 'ladder' construction

Jigsaws

44 Interlocking shapes and careful manipulation of the hot-wire cutter have produced the jigsaw. The tile was painted with water colour paints, and then on the reverse side the interlocking shapes of the jigsaw were drawn and cut

The pupils are able to produce work according to their abilities and ages. A tile can be cut into a few large pieces or many intricate ones.

The only limitations imposed were that the pieces had to interlock and the pupil had to be able to piece together his own jigsaw. The number of pieces that could be produced from one tile proved to be a powerful challenge.

MARION CENTER
HSIMC

Mosaics

In the previous exercises, an accumulation of small pieces of tile have been collected. These can now be put to use. The pieces of tile can be coloured with water-based paints and made into mosaics. This is a suitable subject for project work. (See Plate 1 opposite page 16.)

The tiles have to be mounted onto a rigid board. A sheet of 12 mm ($\frac{1}{2}$ in.) paper composition board is suitable, but has to be strengthened with a wooden framework, glued and screwed to the back. This strong but lightweight base is then given a coat of white emulsion paint on the working surface. (See figure 46, overleaf.)

The selected design is transferred to the board, showing the main lines of the composition. The circles of expanded polystyrene are the most important part of the mosaic and are glued into position, using a heavy-duty wallpaper paste. The connecting and overlapping lines produce the rest of the composition. These areas are then filled in with the sections of coloured tile.

The completed mosaic is finally given a coat of varnish. It is important to test the varnish before application. Some varnishes react adversely upon the tiles.

45 Stained glass window *180 cm × 75 cm (6 ft × 2$\frac{1}{2}$ ft), boy 15. The black lead lines are completed first and then the areas filled in with coloured tiles*

PLATE 2 (opposite) *Jesus stills the water, 150 cm × 110 cm (5 ft × 3$\frac{1}{2}$ ft), boys aged 15*

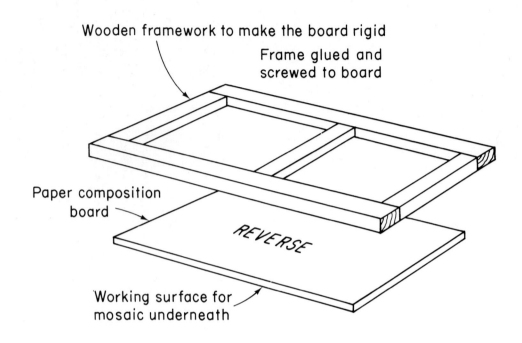

46 *Rigid board for mounting mosaics*

Action of solvents

Tiles can be decorated by the direct use of flame. *Self-extinguishing tiles must be used.* The heat of a candle flame melts the tile, darkens it with deposits of soot from the smoke and leaves holes and an undulating pattern.

47 This light box is constructed of tiles, which have been pinned and glued. Holes have been burnt into three tiles by a candle flame and after these have been covered with pieces of different coloured cellophane, the tiles are placed parallel in the tile box. A light source is concentrated at one end of the box, which is viewed from the opposite end. There is a mixing of colours where the holes overlap in the three tiles

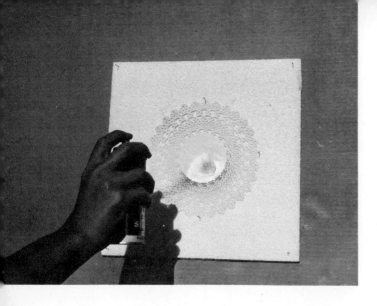

It is possible to produce moving parts for the intermixing of colours.

It has been stated earlier that some adhesives react upon expanded polystyrene foam. By careful, controlled experiments and under the close supervision of a teacher in school, a list of reactors can be made. Nail varnish, some adhesives, cellulose paints, varnish and thinned paint brush cleaners burn into the surface of expanded polystyrene. These properties can be put to creative use, but emphasis must be placed upon safety precautions, and the need for a well ventilated work area.

48 A stencil is pinned to an expanded polystyrene tile. In this example, a paper doily has been used
49 An aerosol of gold or silver lacquer paint spray is used. Hold the spray at the correct distance, according to the maker's instructions, and direct the spray at the tile
50 Upon removal of the stencil, it will be found that the pattern remains, but it will be in relief. The unprotected areas will have been burnt by the reaction of the gold spray upon the polystyrene. Through experience, the amount of paint required to dissolve the surface and to produce the relief will be discovered. Too much paint will dissolve the tile completely

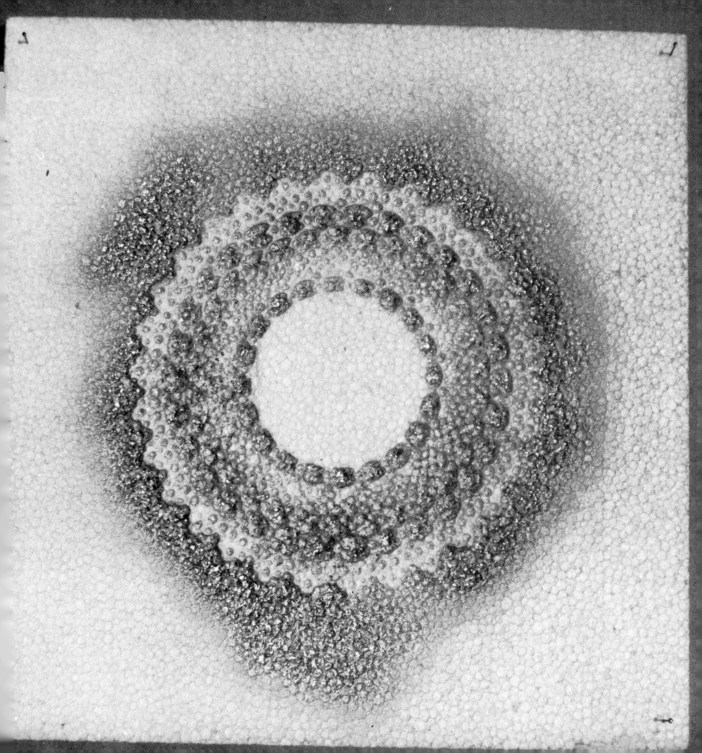

51, 52   Pieces of card are cut in the shapes of cogwheels. They are then pinned to the polystyrene and sprayed with gold paint. Upon removal of the stencils, the tile has a golden background and white cogwheels in relief. The card cogwheels which are also gold are pinned over the existing white reliefs, thus producing three planes

Applied work

Expanded polystyrene is a useful material for applied work. Its strength and lightness make it easy to use for mounting and display. Mounting work on tiles, using steel pins instead of a permanent glue, means that it can be used again. The white purity and texture of the tiles enhances the applied work.

53, 54 A fish *(girl 8)*. *A tile, a sewing needle and a supply of yarn are required for mathematical stitching. The threaded needle is pushed through the tile and by careful planning, designs are produced consisting of straight lines. There are many possibilities using this simple method*

55 Clown *(girl 13). The use of expanded polystyrene tiles as a form of touch tapestry is quick and effective compared to the more traditional methods. The design is drawn on a tile with a ball point or fibre tipped pen and thick pieces of coloured knitting yarn are pushed into the tile. The springiness of the tile surface locks the wool in position. A small screwdriver or blunt piece of wire are excellent as 'pushers'. The outline of the design is followed with the yarn, and then areas of the composition are filled in with coloured yarns*

56 Sunflower

57 Shapes of coloured material are stitched to the tile without using glue. The colour and smoothness of the material is shown to its best advantage against the whiteness and texture of the tile

58 This was designed by a girl aged fourteen, using french knitting and oddments of coloured yarn and string. The materials are fixed with steel pins. The design is emphasised by painting the background in a dark colour as a contrast to the materials and white tile

3 Take a slab . . .

For large pieces of work 12 mm ($\frac{1}{2}$ in.) ceiling tiles are rather fragile. Slabs, up to 100 mm (4 in.) thick, are available from general art suppliers, and slabs thicker than this can be obtained from specialist plastic concerns.

Many of the exercises using tiles are suitable for slabs, with the added thickness giving extra strength.

59 The design for an abstract relief is lightly sketched on the surface of a slab of expanded polystyrene. It is important to decide upon the depth of the relief. A constant depth can be obtained by either fitting a cork on the heated carver, or bending the carver at right angles to the required depth. The outline is followed by the carver, and the inside areas are then removed, by burning or by 'scrumbling' with a blunt instrument

60 The finished relief is painted, to emphasise the design and relief areas

61 Carving taken from a brass rubbing (girl 15)
62 St Francis *Gold powder paint finish (boy 15)*
63 Cormorant *Painted relief (boy 14)*

64 Battle *(boy 15)*. A large slab, 150 cm × 60 cm (5 ft × 2 ft), depicting a battle scene, has been carved using heat cutters. This has been used for large fabric prints for a tapestry

65 St George and the dragon *(boy 15)*

66, 67 A Picasso style face was originally used as a print. The print consisted of two planes. This has been extended into an intricate relief, but still retains the original concept. The textured effect was achieved by rubbing a coloured chalk over the most important planes

PLATE 3 (opposite) *Melted wax study* by the author

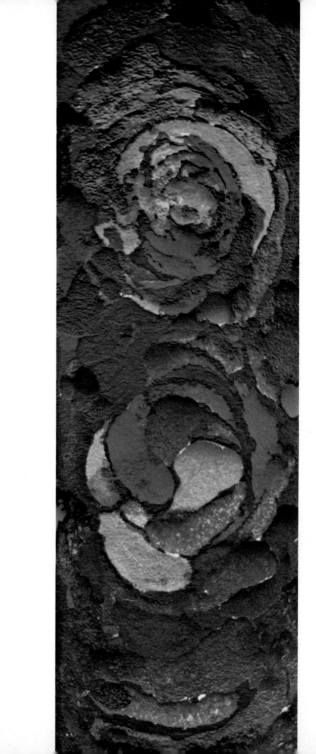

68 A slab, 120 cm × 45 cm (4 ft × 1½ ft), using the design of a Polynesian shield, has been cut using a sharp craft knife. The photograph shows the original relief with a print

69 Spaceman *(boy 14)*

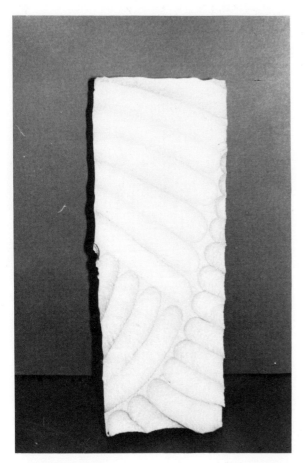 

70 Relief designs can be produced by shaping and carving. Shaped heated cutters can be used, especially if specific shapes are required. This example, an enlargement of a shell, is a useful exercise in controlling the cut and discovering the heat required

71 This slab has been carved using heat and saws; then given a finish using a variety of coloured paints

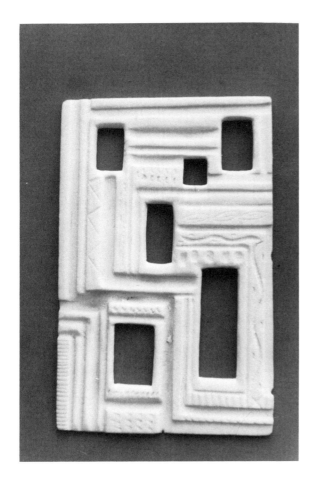 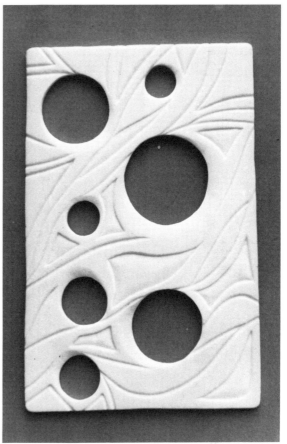

72, 73 The thickness and strength of slabs make expanded polystyrene a suitable medium for using files, saws and wire brushes. These traditional shaping tools produce very pleasing textures. By careful use of planes and contrasts of texture, rounded effects can be obtained from thin slabs, as in these designed by boys aged 14 and 15

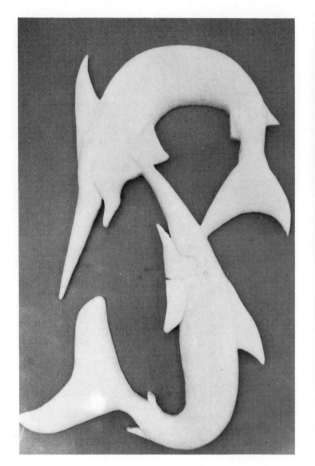 

74 Swordfish *(boy 15)* carved from a 5 cm (2 in.) slab

75 Knight *(boy 15)* using two large slabs, height 180 cm (6 ft)

Melted wax reliefs

Wax crayons may be melted in a test tube by the heat of a candle. When the wax turns to a hot flowing liquid, it is poured onto the surface of a slab. The heated wax melts the polystyrene foam and produces a coloured, heavily sculptured relief. Care must be used in the designing and use of colours. (See figures 76 and 77 on this page.)

It is advisable to experiment with this technique in order to judge the heat and depth required. Some adhesives, varnishes and other liquids react against expanded polystyrene. Great care must be used in all experiments.

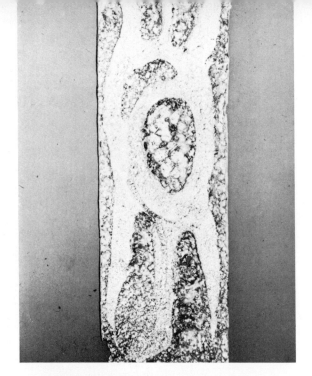

78 A solvent has been painted on the surface. The foam has melted to a variety of depths, and the work has been completed by a fine spray of gold paint

79 A design, etched with nail varnish, has been completed by painting the highest plane.

80 A slab has proved an adaptable material for applied work. Nails and screws of various shapes and sizes are pushed into the slab at different heights and in varying density, to produce a three-dimensional nail mosaic. The initial design of rhythmic curves is outlined using black-headed screws, and the spaces are filled in with nails of contrasting shapes, colours and sizes

81 A large mural for a staircase wall in a large children's shop. The expanded polystyrene foam has been toughened by covering with a thin layer of plaster, and has a painted finish

Carving and casting

The speed of carving expanded polystyrene and the great variety of textures has been recognised architecturally. Modern buildings make great use of concrete slabs, but too much in large areas has been criticised because of its lack of interest. This can be overcome by the use of decorative mural designs, made from casts of expanded polystyrene.

In school, carved expanded polystyrene can be used for casting in clay, plaster or concrete. The work is usually on a small scale because of the difficulties of fixing and mounting.

The technique is simple but effective; it shows the pupils that what is done on a small scale in school is also done on a large scale outside, and will make them more visually aware of their environment.

82 The design is drawn on the slab, and a rough negative carving is produced. An electric router is being used. This is an attachment to an electric drill and is usually used for shaping concrete. However, a circular wire brush fitted to an electric drill will produce the same effect as a router in sculpting the material

83 The final details are carved. The edge border acts as a frame to contain the layer of concrete to be poured upon the carving

84 When the concrete has set, the original expanded polystyrene carving is chipped away from the concrete cast. Small areas are removed with a wire brush

85   An abstract design is carved in a slab of polystyrene foam
86   The carved foam is placed inside timber shuttering ready for the pouring of the concrete
87   Cleaning away the original carving
88   The finished design, showing the fine reproductive detail produced in concrete from the expanded polystyrene cast

89   Expanded polystyrene being carved for a large architectural relief. The clear-cut outline and smooth finish are well illustrated. Notice the type of equipment in use
90   The carved sections are placed in position ready for the shuttering and casting

91 The completed precast concrete mural, consisting of eight sections each 6·6 m × 1·5 m (21½ ft × 5 ft), with a maximum relief of 200 mm (8 in.)

92 This detail of one of the panels shows the variety of textures and quality of finish

93 A richly decorated precast concrete wall, cast in situ from expanded polystyrene

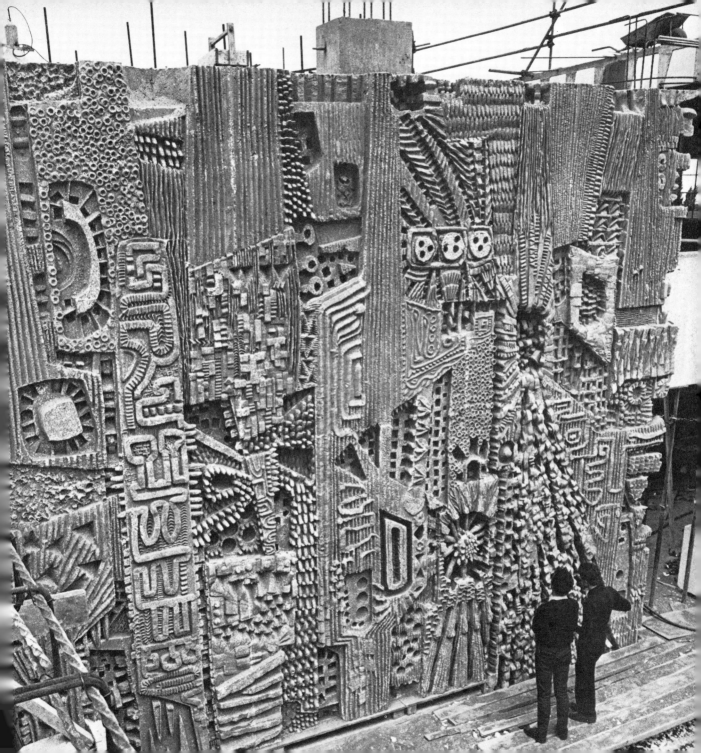

4 Take a block . . .

Large carvings

Expanded polystyrene can be obtained from builders' merchants and art suppliers in large block form, and is very suitable for carving. It does not require expensive equipment or strength to cut, and therefore can be used by the youngest of children. The ease of carving and cutting is a major advantage and pupils are not deterred by technical difficulties. Small pieces of work can serve as maquettes, and larger pieces can be used outside in the open air, so long as they are protected from the weather and attacks by birds. This can be done with plastic emulsions, waterproof plasters, concrete washes or glass fibres.

Carving is the shaping of a material by subtraction, by the careful removal of the substance. It is necessary to become familiar with the nature of the material, its limitations and the equipment required.

94, 95 Small blocks are cut with fretsaw blades and shaped by a variety of files. By experimenting with angled cuts, an infinite variety of forms are produced. After a few cuts a definite pattern begins to emerge. This is the exciting start to carving, a lesson in materials, equipment and creativity

96, 97 An oblong block is carved with heated carvers. The outside edges of the shape remain unaltered; the design depends upon the relationship of the spaces produced. It must be remembered that the finished shape must look effective when viewed from all sides. The horizontal and vertical relationships are enhanced by clean lines and smooth sides

Irregular shapes can be improved by the addition of textured surfaces. The application of extra heat will cause the sides to buckle and produce unusual contortions.

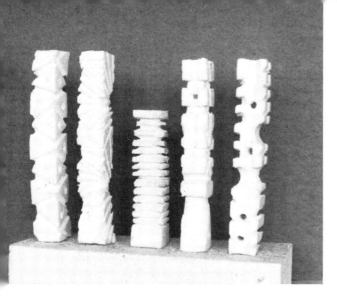

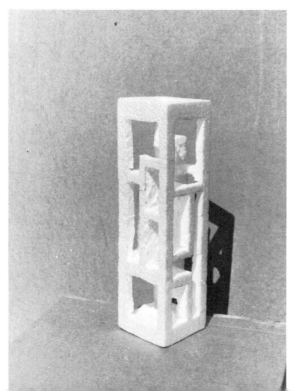

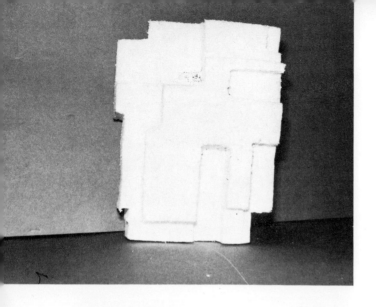

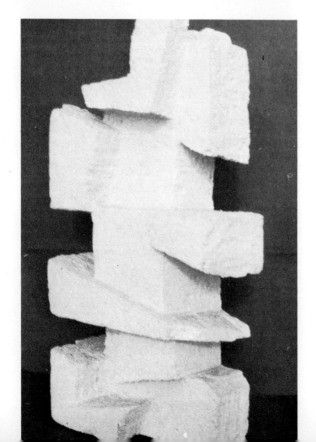

98 A rectangular block is carved into a series of shaped relationships. All the plane surfaces are parallel to the original surfaces of the block. Horizontal and vertical lines have been translated into a three-dimensional form, and a sense of proportion maintained

99 An extension of block carving. The limitation of using only parallel lines has been removed, but emphasis is still placed upon the relationship of plane surfaces

PLATE 4 (opposite) *Scrap monster man,* children aged 9

100 Puppetry is a valuable introduction to the carving of heads and figures. Many traditional methods take a long time and often techniques prove difficult

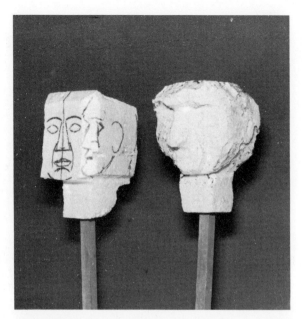

101 The front and side views of a head are outlined on a block of expanded polystyrene. Using heated carvers, or files, the outline is carved. During carving, the head must be continually viewed from all sides

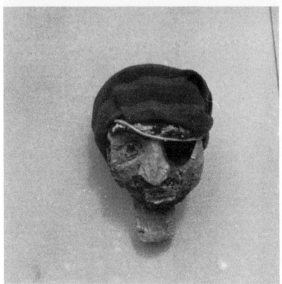

102 The carved head is then painted. The headscarf and the eye patch have been pinned into position to add character. These can be removed later so that the head can be used again for other studies. The hole in the head and neck for manipulation is made by a heated cutter

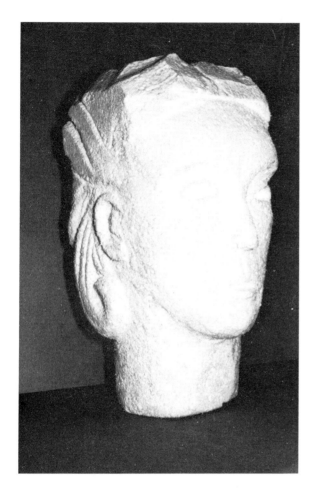 

103 This partially completed head has been carved with files and sandpaper. A more permanent finish is achieved by giving the head a coating of plaster

104 Very detailed carving has produced these imaginative totem poles. They can be produced by a single pupil or as a group activity. Each pupil carves a head and these are then glued together or threaded on a metal rod

105 A group of Nativity figures depicted in simple shapes

106 A carving of Nelson to be mounted on a 16 m (52 ft) high column; Nelson and the column are made of expanded polystyrene. The versatility of the material enables sculpture work to be produced quickly and any finish can be simulated perfectly

107 A copy of a traditional Indian carving, showing the amount of detail possible in expanded polystyrene. To give a metallic appearance, a coating of thin plaster is applied. This is then covered with a metallic paint. This can be toned down by applying a coat of black polish over the paint, and a little rubbing will produce a semi-matt finish

108, 109 Large pieces of work have to be made from multiple blocks of expanded polystyrene. A scale model is produced and the blocks are glued together and then carved

110 The completed 6 m (20 ft) high Trojan Horse carved by Derek Howarth

Few of the millions who watched the investiture of HRH The Prince of Wales realised that the giant Coat of Arms on the wall of Caernarvon Castle was carved from expanded polystyrene. (See figure 111, on previous page.)

112 Full scale drawings were made, transferred to the polystyrene and cut by hot wire. Working with saws and a wire brush, sculptor Derek Howarth produced this 5 m × 3 m (15 ft × 10 ft) Coat of Arms. This was then coloured before it was covered in glass fibre (see figure 111)

113 A polystyrene mural 20 m × 4 m (65 ft × 13 ft) in a London restaurant. The relief is emphasised by the careful use of coloured lighting

114 Railway wagon *3 m × 2 m × 1 m (10 ft × 6½ ft × 3 ft), showing detailed carving and suspended from the ceiling of a hotel*

115 A 7 m (23 ft) dragon carved from multiple blocks of expanded polystyrene. The surface has been reinforced with plaster and then coloured. The carving, although very large, shows clearly the great saving in weight of expanded polystyrene over traditional materials, its strength and the speed and ease of installation

Sparklets Beertap

Sparklets Beertap

Sparklets New Hostmaster

Sparklets Corkmaster

116 Flexibility in design, light weight and quick erection makes polystyrene foam a highly suitable material for exhibition and display. The exhibits illustrated above were made in toughened polystyrene foam, and show the remarkable degree of finish which helps the exhibition designer achieve an eye-catching feature

117, 118 These enlarged items show that expanded polystyrene can be carved with great accuracy. The spoon complemented an enormous cup and saucer made for advertising purposes

5 Take some scrap . . .

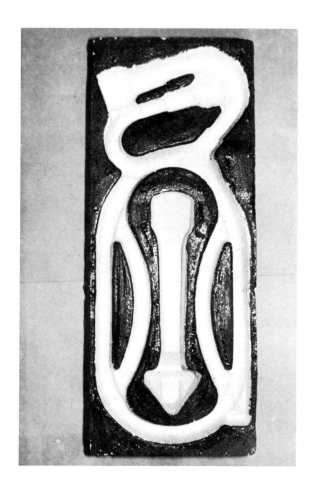

Expanded polystyrene can be most easily obtained in the form of protective packing. Electrical equipment, fruit and glass and many other commodities use preformed polystyrene. From these scraps, many exciting pieces of work can be created, and they can be used in many of the exercises mentioned in earlier chapters.

119 Scrap
120 Polystyrene, previously used to pack a shower unit, makes a very interesting relief, emphasised by the use of paints

The following figures show a selection of work by young children using scraps and oddments of polystyrene packing.

121 The addition of shells has made an excellent mouse (boy 7)

122 *A selection of work by children aged 8*

123 A many-funnelled ship (boy 8)

124 Indian head *(boy 10)*

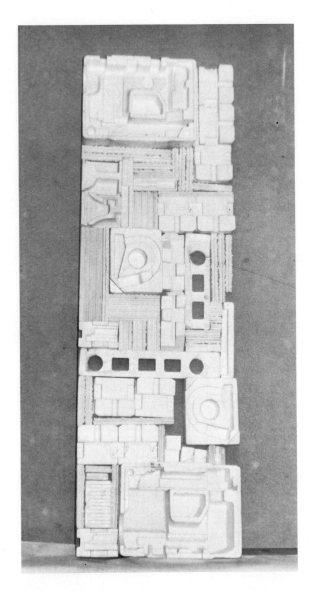

125 A selection of shaped packing cases have been fastened to a baseboard to form an interesting relief

There are many variations on the relief theme. These are some suggestions: A notice or display board made by incorporating notices in the hollows of the relief (see figure 126). Various areas within the relief can be painted, using a limited colour scheme. A plaster or concrete case can be taken, thereby making a permanent decorative mural.

Junior ART CLUB MONDAY 4 to 5 p.m.

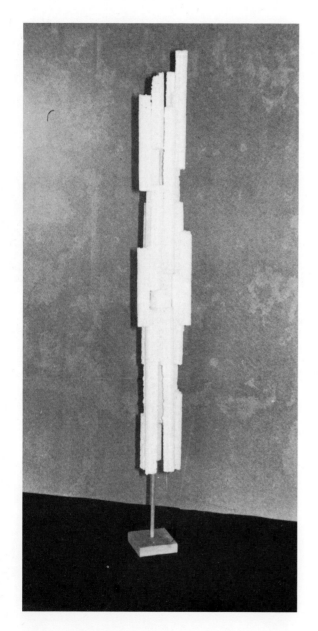

127 Construction *using pieces of scrap polystyrene originally used for the packing of television valves*
128 Polystyrene spheres and cocktail sticks make this atomic study

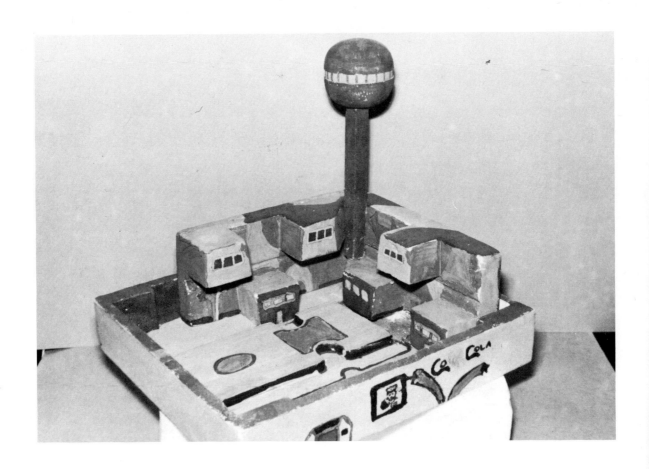

129 City of the future *consists of two packing cases and a hollow sphere that had contained a small perfume bottle. No carving was necessary. The parts were painted after gluing, giving this futuristic concept of a self-contained community*

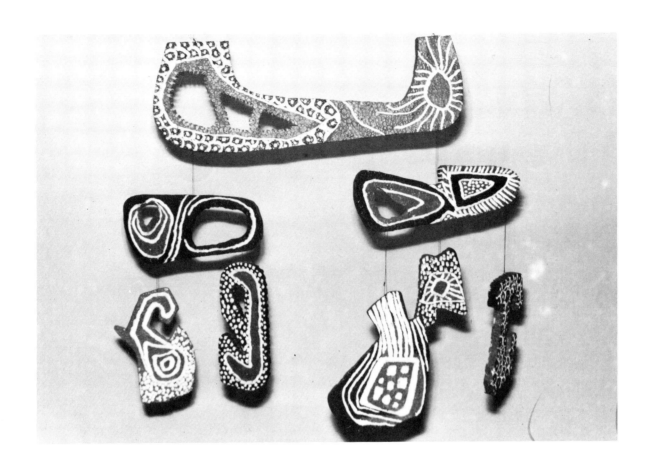

130 Pieces of scrap polystyrene have been carved, the surfaces etched and textured with abstract designs, and then painted using two contrasting colours. The material, being very light in weight, is suitable for making mobiles

Further reading

Creative Drawing   Ernst Röttger and Dieter Klante

Creative Paper Craft   Ernst Röttger

Creative Wood Craft   Ernst Röttger

Creative Clay Craft   Ernst Röttger

All published by B T Batsford Limited London and Van Nostrand Reinhold New York

Hot-wire machine   P J Clarke, Allman London

Plastics as an Art Form   Thelma R Newman, Pitman London

Suppliers in Great Britain

Expanded polystyrene

E J Arnold Ltd (School Suppliers)
Butterly Street
Leeds LS10 1AX
(also PVA adhesive)

Baxendale Chemicals
Nr Accrington
Lancashire

Elford Plastics *(polystyrene spheres)*
Elland
Yorkshire

Magros, Division of Eagle Pencil Company
Woking
Surrey
(also Marvin Medium, battery cutters and electric carvers)

Venesta International – Vencel Ltd
Erith
Kent

Hot-wire cutters

Super Tools (1951) Ltd
Shepherds Bush
London W12

Adhesives

Tretobond Adhesives
The Hyde
London NW9

Suppliers in USA

In the US, two well-known brand names for expanded polystyrene are *Dylite* and *Styrofoam*.

Styrofoam is available in sheets or blocks from most well-stocked art supply stores. It can also be obtained from many novelty or party supply stores.

Of course, as has been mentioned in the text, anytime that you buy an appliance, Christmas decorations, or anything fragile, you are more than likely to find yourself with a plentiful supply of expanded polystyrene packing material.